Training for decentralized planning

Lessons from experience

Vol. 1

by
Materne Maetz
and
Maria G. Quieti
Development Policy Studies
and Training Service
FAO Policy Analysis Division

TRAINING MATERIALS FOR AGRICULTURAL PLANNING

29

FOOD AND AGRICULTURE ORGANIZATION OF THE UNITED NATIONS
Rome, 1987

Reprinted 1993

The designations employed and the presentation of material in this publication do not imply the expression of any opinion whatsoever on the part of the Food and Agriculture Organization of the United Nations concerning the legal status of any country, territory, city or area or of its authorities, or concerning the delimitation of its frontiers or boundaries.

M-67
ISBN 92-5-103420-6

All rights reserved. No part of this publication may be reproduced, stored in a retrieval system, or transmitted in any form or by any means, electronic, mechanical, photocopying or otherwise, without the prior permission of the copyright owner. Applications for such permission, with a statement of the purpose and extent of the reproduction, should be addressed to the Director, Publications Division, Food and Agriculture Organization of the United Nations, Viale delle Terme di Caracalla, 00100 Rome, Italy.

© **FAO 1987**

Foreword

The shift of emphasis to growth-with-equity policies and the disillusionment with central planning have contributed to the growing interest among developing countries in decentralizing some planning and management functions. Decentralization, referred to in this paper as an overall institutional effort to decentralize decision-making according to a process nationally defined, is undertaken in order to promote a more balanced development, to boost mobilization of local resources, to strengthen local political institutions and people's organizations. The creation or strengthening of decentralized structures for decision-making, however, does not by itself guarantee that there will be more economic growth, greater social equity and greater efficiency in the delivery of public goods and services. New essential skills and attitudes among planners are called for in order to ensure these expected benefits from decentralization through planning practices that reflect better the empirical realities within a country.

Experiences drawn from a number of countries show that in-service training can be a means to assist in creating such planning capability through upgrading the skills of planners and inducing behavioural changes aimed at enhancing developmental rather than mere regulatory functions. When coupled with technical assistance, in-service training can also help bring forth improvements in planning methods and procedures.

This paper, accompanied by selected illustrative case studies, reviews a number of decentralization experiences with the training programmes that were elaborated to prepare planners to the expected new tasks. The review has been made with the help of a framework which is also proposed as a tool to analyse decentralization for trainers facing the task of designing in-service training programmes. A number of issues for consideration are highlighted which should be of interest to professional trainers from academic/training institutions, policy-makers involved in the planning and implementation of decentralization policies, regional institutions and international agencies working in support of decentralization and training. The paper should be considered as an attempt to gather and disseminate information as well as to raise issues for further work and investigations. It is in this light that we would welcome any comments, observations and additional information that may lead to improved practices in planning and management of development activities.

Howard W. Hjort
Director
Policy Analysis Division

CONTENTS

VOLUME I

	Page
INTRODUCTION	1
PART I. ANALYSIS OF DECENTRALIZATION AND IDENTIFICATION OF TRAINING NEEDS	5
1. A framework to analyse decentralization	5
1.1 Administration/Organization	7
1.2 Finance	10
1.3 Planning	12
1.4 People's participation	16
2. Types of decentralization	18
3. Identification of training needs	26
PART II. TRAINING FOR DECENTRALIZED PLANNING: SELECTED ISSUES	32
1. Timing of the training programme	33
2. Participants in the training programme	37
3. The content of the training programme	38
4. Institutional set-up for training	45
5. The need for a country specific approach	48
SUMMARY AND CONCLUSIONS	50
References	55

ANNEX 1. Framework for the analysis of decentralization

VOLUME II

CASE STUDIES

No. 1	INDIA	Association of Voluntary Agencies for Rural Development (AVARD), by S.D. Thapar
No. 2	BANGLADESH	Academies for Rural Development, Kotbari, Comilla, by Md. Abdul Quddus
No. 3	INDIA	Institute of Social and Economic Change (ISEC), Bangalore, by G. Thimmaiah
No. 4	NEPAL	Agricultural Projects Services Centre (APROSC)/F.A.O, by M. Maetz
No. 5	NIGER	Ministry of Planning/F.A.O., by M.G. Quieti
No. 6	SRI LANKA	Several Training Institutions, by M. Perera
No. 7	ZAMBIA	Provincial Planning Unit, Southern Province/F.A.O., by C.H.B. Muleya and D.C. Mbilikita
No. 8	\multicolumn{2}{l}{Institut Panafricain pour le Développement (IPD), Ouagadougou, Burkina Faso}
No. 9	\multicolumn{2}{l}{United Nations Centre for Regional Development (UNCRD), Nagoya, Japan, by K.V. Sundaram}

INTRODUCTION

There is a growing interest in decentralized planning and implementation among developing countries throughout the world. Transfer of planning and administrative authority from the central government to sub-national levels - governmental or non-governmental units, administrative or political institutions - has already been undertaken in very diversified ways and with uneven degrees of success. Yet, more countries are envisaging to decentralize. Generally, the objectives of decentralization as put forward by the countries are such as:

- promotion of a more balanced development in the country;

- design of more realistic projects and programmes which take into account local potentials and constraints;

- more effective coordination of development activities at various spatial levels through disaggregation of planning functions;

- strengthening of local political institutions and increase of people's participation in development;

- boosted mobilization of local resources.

In this text, **decentralized planning** is understood as the result of an overall institutional effort undertaken by a given country to decentralize decision-making according to a process defined throughout the country. Accordingly, this concept differs from **local level planning** which appears more as a result of local, often isolated, initiatives, without special institutional set-up or links with the national level.

Part of the overall institutional effort is the building up of a capability at central and regional level for planning and implementation. This is a necessary prerequisite to decentralization and in fact, in some cases, it is deemed that the assessment of such capability should determine to a great extent the scope of decentralization and the allocation of functions. Capability building is here referred to as encompassing not only skilled personnel but also its management and the methods and procedures characterizing its work environment.

In-service training is generally recognized as a necessary input of decentralization programmes. Coupled with technical assistance it can bring forth improvements in planning methods and procedures. Primarily, however, it prepares the people involved to face changes in duties and responsibilities through upgrading of their skills and the introduction of behavioural changes, commonly felt requirements in many countries attempting to decentralize throughout the world.

In **Asia** decentralization has long been in existence and its implementation has taken different forms. The creation of sub-national units of administration, like the provinces in Thailand and the Philippines, has not always meant the empowering of such units with real planning functions. Room for planning and for participation of local organizations can be witnessed in Nepal, India, Sri Lanka and Bangladesh at the district level, however restricted the scope of planning may be in some cases. Among the factors cited as hindering the process of planning and implementation at sub-national level are the lack of experience of staff in working in multi-disciplinary teams, lack of appropriate methodologies and techniques, as well as the absence of regular and frequent interchange of staff between the central and the sub-national planning units (1).

In some countries of **Latin America**, like Brazil, Argentina, Venezuela, Mexico, the decentralization of planning and implementation has been carried out through semi-autonomous organizations with mixed results. Most notable has been the difficulty of reconciling regional with national priorities and of associating not only people's organizations but also government officials to the regional plan formulation. The issues of establishing adequate institutional linkages between the centre and the regions and of building up human resources both from the government side and the population are a recurrent theme in the literature on decentralization experiences in Latin America (2).

In the **Near East and North Africa,** the dearth of qualified and motivated staff is considered as one of the serious constraints to the regional decentralization initiated in the 70's. Although advocated politically, the decentralization process is fraught with difficulties in practice, ranging from inadequate financial devolution to sub-national units to weak coordination mechanisms between centre and region for the collection, storage, processing and retrieval of data and information. Also weak are the procedures for the regionalization of the plan and the budget (3).

In **Africa South of Sahara** the decentralization of planning and management was initiated in the 60's and 70's and it is a policy being officially pursued by a large number of countries nowadays. This has called for the creation or strengthening of sub-national units with planning and management functions and the creation or reviving of people's organizations. Redeployment of staff at sub-national level and their training with donors' assistance seem to be a constant feature of the decentralization programmes although the lack of coordination, the weak coherence and pertinence of much of the large amount of training available to Africans, within and outside the continent, are pointed out as areas for remedial action (4).

It is recognized that there is no unique or ideal set-up for decentralized planning. Each country has its specificities and attempts to decentralize differ according to national conditions, ranging from simple disaggregation of the national plan to genuine participation of the population in the planning process and using different institutional arrangements (5). Similarly, there is no unique or ideal approach in training for decentralization. A great variety can in fact be found in the training programmes elaborated in support of decentralization by national and international institutions.

FAO has been engaged in a number of countries with training-cum-technical assistance projects since 1976 (e.g. Nepal, Morocco, Tunisia, Zambia, Colombia, Brazil). The Commonwealth Secretariat has also been engaged in this area and has recently held a seminar in Zimbabwe on training for Decentralized Administration in Eastern and Southern Africa. The Asian Institute of Technology, Bangkok, Thailand, has created a regional network of institutions involved in training for regional/sub-national planning. Many other training programmes exist, some of them with very interesting and original aproaches. Up to now little has been done to gather and share such experiences. National training institutions often work in isolation, without benefitting from lessons learned in other countries (or even within their own country).

This paper intends to review a number of experiences by developing countries in designing and running training programmes for decentralized planning with the aim of disseminating such information, drawing operational conclusions with implications for trainers, policy makers and lending/donor institutions. It also aims at providing a methodology to trainers facing the challenge of designing a training programme adapted to their country.

The basic assumption underlying this text is that proper design of training necessitates close analysis and understanding of the decentralization process. For that an **analytical tool is proposed** and illustrated in detail in Part I, Chapter 1. This tool, referred to in the text as "framework", should assist trainers to analyse the process of decentralization and to derive an assessment of the training requirements.

It is by using this tool that **different experiences of decentralization and training are reviewed** in Chapter 2. They refer to Africa and Asia and Pacific regions, as it is in these regions that the trend towards decentralization has been the strongest: Asian examples often relate to experiences having started since the 60's or 70's; African examples illustrate more recent attempts to decentralize. They have been selected so as to depict a large range of both decentralization programmes and training approaches and allow the analysis of the relationship existing between the two. African examples are from Niger and Zambia where two FAO training projects are currently implemented in collaboration with planning institutions. Asian examples are from India, Bangladesh, Sri Lanka and Nepal, where training is provided by specialized institutions in a variety of contexts. They are complemented by two regional training programmes, one in Africa (Panafrican Institute for Development, Ouagadougou) and one in Asia (United Nations Centre for Regional Development, Nagoya). A full account of these training programmes can be found in the case studies gathered under Volume II.

Chapter 3 puts forth a **method for the identification of training needs**. Part II **reviews the major issues identified** and worthy of consideration for the design of training for decentralized planning, like the timing of the programme, the identification and selection of participants, the elaboration of content, the institutional set up, the desirability of country-specific programmes.

The conclusions in the final Chapter draw on the issues raised and point to the need of designing training programmes carefully on the basis of a systematic assessment of decentralization and in particular of the changes brought about in trainees' responsibilities. This will almost inevitably result in nationally-based, job-oriented training comprising specific tailor-made inputs, often requiring the elaboration of an adapted planning methodology or the revision of certain planning procedures. The programmes designed would, therefore, not only have the primary objective of training but also serve as a laboratory to try out a new planning methodology or test improved planning procedures on a limited scale. They would thus provide feedback to central decision-makers from those working at sub-national level on how decentralization is actually being implemented. Concluding remarks on the implications for training institutions, policy-makers and international lending and technical agencies are made leaving a number of issues open for further work and investigations.

PART I. ANALYSIS OF DECENTRALIZATION AND IDENTIFICATION OF TRAINING NEEDS

1. A FRAMEWORK TO ANALYSE DECENTRALIZATION

Traditionally, the great diversity of decentralization experiences is analysed in terms of deconcentration, delegation, devolution and transfer of functions from government to non-governmental institutions. In a nutshell, **deconcentration** is "the geographical redistribution or dispersal of central government administrative responsibilities" (6). There is more or less deconcentration according to how much decision-making authority is given to the sub-national representatives of the central government. **Delegation** implies giving "planning and management functions in respect to specific tasks or projects to organizations which, although funded by central government, do not come under its operational control. Such parastatal bodies are semi-independent and are frequently located outside the normal structures of government" (6). **Devolution** involves the creation of autonomous and independent local government units on which central government has little control and which are established by a law and have the power to raise taxes. This last way of decentralizing is usually considered as "the purest form of decentralization" (6). **Transfer of functions** from government to non-governmental institutions is the "handover from government of some planning and management powers in respect to public functions to voluntary and private non-government organization" (6). In reality, however, decentralization is most often a combination of deconcentration, delegation, devolution and transfer of functions to non-government institutions.

Whereas this way of considering decentralization is of conceptual interest and generally employed in public administration literature, a more detailed analysis is required by those who have to design a training programme to support the decentralization effort. They will need to examine responsibilities given and tasks to be carried out by all those involved in development planning and administration, to evaluate the strength of the commitment of central government to decentralization and the stage of its implementation. The result of this work can then be used to identify the training needs. Responsibilities and tasks will by and large determine the content of a training programme, the government commitment will affect the budget available for training and, consequently, the possible size of the programme as well as the likeliness of involving in it a number of coordinated training institutions. The stage of development of decentralization will dictate the type of training approach to be used (e.g. orientation, skills development, action research, etc.).

In order to conduct such a detailed analysis in a systematic way, a framework has been elaborated, structured so as to answer the following four questions: What type of decentralization ? At what stage of development is it ? What is the extent of government commitment to decentralization ? What are the responsibilities of the major development actors under decentralization?

The framework is divided into four parts, namely:

  i) Administration/Organization,
 ii) Finance,
iii) Planning, and
 iv) People's Participation,

in line with what is felt are four major preconditions of real decentralization : (a) government commitment to decentralization and some deconcentration, (b) some financial autonomy of the sub-national levels both in resource mobilization and decision making on its allocation, (c) clear demarcation and delegation of planning responsibilities to the lower levels and (d) existence of a local political backing to sub-national planning, without which sub-national proposals have very few chances of standing the arbitration stage and of being translated into action.

Each of the four parts of the framework has its specific purpose and provides information to those in charge of designing a training programme for decentralization. In part (i) the extent of deconcentration and delegation is assessed at a general level, along with government commitment and the stage of evolution reached by decentralization. In part (ii) financial devolution and delegation are analysed. These give further indications on government commitment and on possible training objectives. In part (iii) the degree of complexity and deconcentration of the planning process is assessed in order to prepare the ground for an evaluation of training needs. Part (iv) deals specifically with people's participation, one of the most currently declared objectives of decentralization. Its purpose is to find out the extent of devolution to local bodies and/or of transfer of government responsibilities to private or non-governmental organizations and consequently the extent to which there is a need to train non-officials. Each of the four parts is divided in points or questions which have been arranged in a sequence of increasing decentralization. The No/Yes columns are meant to help define each point or question, which can then be elaborated further under "Remarks". The full framework can be found in the Annex.

1.1 Administration/Organization

This part aims at describing the administrative, political, legal will and efforts undertaken to promote decentralized planning. It helps to assess in a preliminary way the conditions in which decentralization is developing. What is the actual commitment of the Government ? The strength with which it is enforcing decentralization ? The perenniality of this policy orientation ? How much deconcentration will there be ? How much delegation ? How likely is real decentralized planning that will consider specific sub-national conditions and needs ? What will be the expected commitment of sub-national staff to the local level, its accountability towards it and its latitude for decisions vis-a-vis the central ministries ? And lastly, what stage of development has decentralization reached ?

This part comprises seven questions which will help to decide whether conditions are there for a training intervention or not, and with what content.

ITEM	NO	YES	REMARKS
1. ADMINISTRATION/ORGANISATION			
Creation of a local planning capability			
Creation of a local planning cell (pluridisciplinary)			
Decentralized planning and implementation procedures developed			
Central level high power committee to coordinate decentralization			
High calibre staff transferred to local level			
Accountability of government staff to local level authority			
Staff performance assessment made at local level			

Decentralization requires to give decision-making authority for planning to sub-national levels which so far have only had implementation responsibilities. If this is to take place, the government will have to take action in order to create a **local planning capability**, be it governmental or non-governmental, by creation of specialized planning units/bodies or by training and upgrading existing sub-national implementing agencies. Such action will be in terms of circulars or orders emitted by the government, training, changes brought in staffing pattern, etc.

The creation of a **planning cell**, at sub-national level, is usually considered as the most radical step to create a decentralized planning capability. In case such step is taken, it should be made clear whether this cell will have the responsibility for overall planning or only for specific aspects (physical planning, agriculture, rural development, etc.), and the kind of staffing envisaged or existing should be specified (generalists, specialists, multidisciplinary team). This information will evidently bear on the potential content and participants in a training programme for decentralization.

In certain countries, although the will to decentralize and the planning infrastructure exist, the **procedures for decentralized planning** are not specified. This is a great constraint to effective decentralized planning. Precision on whether procedures exist, are under preparation or not, gives indication of the stage decentralization has reached. These procedures should pertain to the definition of the actions involved in planning, their planning responsibilities, the outputs they have to produce and the planning relationships existing among them. In case procedures exist, the way by which the new rules are enforced should be mentioned: administrative circulars or rules, government orders, laws or acts of parliament, constitution. This will help to assess the stability of decentralized planning: can the planning process be changed easily at any time or are there lengthy or complex procedures to protect it ? Description of planning procedures will also be an essential part of the training content.

Even though procedures exist, they may or may not be followed. If followed, they may be so in spite of the spirit of decentralization. These problems often arise because of the reluctance of central ministries and departments towards decentralization, which, they rightly think, will make them lose some of their power. In some countries, a **special powerful central level committee** is set up to coordinate the ministries and departments as well as monitor and direct the change of their working procedures into decentralized procedures. If such a committee exists, its duties and authority should be studied. It could eventually be given the responsibility of coordinating the various institutions involved in a national training programme for decentralization.

It is sometimes observed that although a wide range of planning responsibilities have been handed over from national to sub-national levels, no **transfer of high calibre staff** has taken place from the centre to the regions to improve local capabilities and increase local coordination of planning. This situation is quite illogical and a constraint to the success of decentralization. It will call for additional training efforts to raise existing staff to the level required by their new responsibilities. In case such transfers did take place, it would be useful to list what action the government took to make them effective: by authority, by financial or career incentives, by establishment of additional office and lodging facilities, etc.

If decentralization is to give more responsibility and authority to the sub-national levels, it requires also that government staff working at this level be made **accountable administratively and financially** to the sub-national level authority, be it a chief official (prefect, governor, executive secretary) or planning authority (council, planning body), the technical accountability continuing eventually to be towards the centre. This is very rarely achieved because of the resistance of the centre, the result being that in certain cases a dual administration can be found locally: one accountable to the central ministries and departments, and one fully accountable to the local authority. Such cases should be carefully analysed as this will be determinant information for selecting participants and deciding on the training content.

Financial and administrative accountability of sub-national staff to local authority will certainly depend largely on the **administrative authority** given to it. It may be necessary to list the authority given to the sub-national level in order to assess whether accountability will be effective. Items to be considered are amongst others: appointment, dismissal, disciplinary measures, staff assessment, promotions, transfers, payment of salaries and allowances, selection for training, etc. It may be useful to analyse how career development is managed and by whom. This analysis will be important when approval of training programmes, nomination of participants, mechanisms to motivate trainees, etc., will have to be secured, and will help to find out with which level training organizers will have to deal in priority.

1.2 Finance

The analysis of financial rules under decentralization helps to assess the extent to which sub-national levels are given the opportunity and latitude to take decisions in development planning and implementation. It helps to verify whether the sub-national levels are really given the resources and the liberty to use them in line with what may be stated in policy documents on decentralization published by the government. The part of resources devolved to the local levels, the degree of flexibility of their use, the encouragement given to them to increase their autonomy are good yardsticks for measurement. It also helps to detail some specific tasks to be implemented locally.

This part comprises seven questions.

ITEM	NO	YES	REMARKS
2. FINANCE			
Resources earmarked for decentralization			
Funds allocated to local level			
Criteria to allocate funds to local level			
Existence of un-tied funds			
Power to levy taxes given to local authority			
Incentives provided for local resources mobilization			
Financial accountability of government officers to local authority			

Decentralization to be successful needs a **specific budget** for its promotion and operation. The public needs to be informed, campaigns have to be organized, training must be run and documents and circulars printed and distributed. Administration often requires to be strengthened in manpower and supplies (stationery especially), new constructions need to be made. This indication will be valuable while deciding the possible scope and size of a training programme.

Decentralization is also expected to have an impact on how the national budget is prepared, structured and managed, in particular on how **funds will be allocated to sub-national levels.** The questions are: what proportion of the budget is allocated to the sub-national levels (in percentage), and what is the allocation mechanism ? In case the allocation is made with the use of a formula, criteria used should be analysed. This helps to assess the latitude given to the sub-national levels. The more money goes to them, the more responsibility they will have. If the money is allocated in a transparent and mechanical way, the centre will have less means of exerting pressure onto the local levels in order to direct its use. Management of money flows at sub-national level may require new skills which will have to be provided through training.

Basically, there are two types of funds which may be allocated: **tied funds and un-tied funds.** Tied funds have been pre-allocated sectorally by the centre; un-tied funds are free to be used in any sector, the decision resting with the sub-national level. Tied funds can be more or less strictly pre-allocated: in certain countries, the local level is authorized to reallocate sectorally a certain percentage of the tied funds. This point is very important, and for two reasons: (i) the importance of un-tied funds or the latitude to reallocate tied funds gives a measure of the actual power the centre has conceded, (ii) it helps to check whether sub-national levels will have to make decisions of resource allocation or not. This must be known while designing a training programme, as this new job requires specific skills which will have to be imparted.

As already mentioned, devolution involves giving to the local governments the **power to raise taxes** and mobilize local resources. Has this power been granted ? How much resources can the sub-national level tap in this way compared to what it gets from the centre ? This will give an idea of the degree of autonomy and delimit a new area of potential training needs (resource estimation and mobilization).

Does the government **encourage the mobilization of local resources?** In what way ? By way of providing a matching grant, by incorporating locally-mobilized resources as a criterion while allocating central resources ? By allowing the local level to launch income-generating activities financed through loans ? etc. These points should be analysed to find out the real will of the central government with regard to decentralization. Trainees will have to be informed on them and those concerned will have to be acquainted with specific aspects of design and management of income-generating activities, a field in which, as civil servants, they are likely to have little experience.

With regard to finance, another important point to consider while assessing decentralization is the mode of disbursement of the funds and the **financial accountability**. It is evident for example that if funds transit through a local level treasury to which local level officers are made financially accountable, there will be more decentralization and relative autonomy than if funds were made available by central ministries towards which the accountability would then be. The former case would, of course, require well trained local treasury officers.

1.3 Planning

Once the general environment and the financial set-up have been considered, it is time to concentrate on planning: its procedures, its scope and the delimitation of planning responsibilities at various levels. This is the key part of the framework to be used for identification of training needs and content. It allows to assess the degree of complexity and deconcentration in the planning process. It is possible to evaluate the part of actual planning done by the sub-national levels and to identify the people who have key roles in the process and who, consequently, should be the target group for further training needs evaluation and finally for training.

A review of country experiences of decentralization shows that one of the prerequisites of successful decentralized planning is a strict **demarcation of the planning responsibilities** between national and different sub-national levels. The lack of such distinctions brings almost always confusion into planning and arouses conflicts which are as many constraints to successful decentralization. This demarcation should be clear, detailed and well publicised to all concerned [1]. It is worth documenting it, as it defines potential training areas and target groups. It will also have to be presented during the training programme so that participants are briefed on what is expected from them.

---

[1] A comparison of resources allocated to development activities to be planned for locally, before and after enforcement of decentralization can be a good indicator of the real commitment of government to decentralization.

ITEM	NO	YES	REMARKS
3. PLANNING			
Scope and content of local level planning defined			
Attempt to train and/or provide technical backstopping to local planners			
Local planning authority decides at least on location of projects			
Sectoral targets for the local level exist			
Sectoral targets prepared at local level			
Targets organized and presented in a sectoral plan (standard format)			
Existence of a data base at local level			
Existence of a local level M. & E. cell			
Possibilities of intersectoral planning			
Partially integrated multi-sectoral planning at local level			
Fully integrated multisectoral planning at local level			
Projects formulated at the local level			
Village plans formulated at village level			
Intra-district balance consideration			
Possibility to modify content and costs of projects to a certain extent during implementation without referring to the national level			

The assessment of the amount of **sub-national planning** taking place and the support given to it by the centre needs to be reviewed in many different aspects. Training efforts and existence of eventual technical assistance provided by the centre (central planners, planning centres or institutes, consultant firms, universities, etc.) to the local level clearly give an idea of the support. The amount of planning going on can be judged by a series of simple questions:

- does the sub-national level at least decide on the **location of the projects** to be implemented within its administrative boundaries ?

- do **sub-national sectoral targets** exist ? like for example production targets, training and extension targets, etc.

- are **sectoral targets prepared at the local level** ? in which sector ? prepared by whom ?

- are these targets presented and organized in a **sectoral plan** ?

- does this sectoral plan have a **standard format** defined throughout the country ?

- is there a **data base** at sub-national level, global or sectoral, in which sector ? Is its content and format defined nationally ? How is it used in the planning and monitoring process ?

The answer to each of these questions has implications on a possible training programme and its content. It will determine whether the training will have to cover topics like project planning, target/programme planning, plan formulation, data base establishment and management or not. Planning formats should be carefully analysed as they will have to be explained during the training programmes and trainees should be provided with tools to enable them to fill them correctly.

Good decentralized planning requires that some **Monitoring and Evaluation** (M & E) be conducted at sub-national level. It is worth exploring whether a specific M & E unit has been established and what its staffing pattern is. This unit will be a potential target for training.

In the case of many countries, one of the objectives of decentralization is to improve **intersectoral coordination**. In this regard, the planning process should be analysed to see whether possibilities of intersectoral planning are allowed by special planning mechanisms like sectoral and intersectoral planning committees. In certain conditions it may be feasible to promote integrated multisectoral planning, that is planning where (a) a priori analysis of the global situation of the concerned areas is conducted, (b) there are some funds which are not a priori earmarked for a particular sector or alternatively there is some flexibility in the use of the sectoral funds, (c) intersectoral linkages are considered. The proportion of the total outlay planned for in this way at the sub-national level should be determined and will be a criterion to decide whether specific training programmes should be conducted on this topic, and the context in which integrated multi-sectoral planning can take place will determine the approaches or tools that can be disseminated through training.

Decentralized planning may encompass **programme planning and project planning**. In case projects are planned at sub-national levels, distinction should be made between normative planning, where planners can choose between a series of projects pre-planned technically and financially, and real planning where local planners formulate original projects and make specific analyses. According to the situation observed, the content of the training to be provided will be quite different.

There is a great diversity in the decentralized planning set-up used by different countries. The planning levels vary in number and so does the output of their planning exercise: it can be project ideas, targets, projects, programmes or plans. It seems necessary to determine the exact **planning output** at each level, on the one hand for training purposes and on the other to determine the integrative quality of the set-up. The more numerous are the levels that will have to come up with plans, the more planning is likely to be integrated, as a plan will compile, and hopefully consider the interactions among a wide range of projects and programmes formulated in detail, technically, financially and institutionally.

Another objective of decentralization which is often put forward by its supporters, is to create conditions for **a more balanced and harmonious development**. One aspect of this objective has already been considered previously in this text when mentioning the need for a special powerful central level committee. But it is also necessary to check whether this same concern is present within the sub-national level and if provisions have been made to take care of intra-provincial or intra-district balance. Accordingly, training will have to cater for this aspect of planning.

The plan provides the guide for development activities to be undertaken, but sometimes conditions require to take some liberty with what was planned. The extent to which the sub-national level can do this **without referring to the centre**, is a good indication of the authority handed over to it. The change to be brought in may be on the content or the cost of an activity or a set of activities.

1.4 People's participation

People's participation may take place at any stage of the "project cycle". It can happen in different ways: it may be a direct participation of each individual through attending meetings or contributing to working teams, or an indirect participation through political bodies - elected or non-elected - (councils, parties, ...), associations, groups or cooperatives. For example, identification of project ideas may be done through a village meeting or through a survey of selected individuals; preparation of a project may be handled by a group of elected representatives working together with technical officers; project appraisal and selection may be the responsibility of a council; implementation may require donations or voluntary work by citizens, etc. According to the participatory approach envisaged and its importance, it may be necessary to inform or train those who have to play the most important roles in it.

In most cases, under decentralization, planning responsibility at sub-national level is given to an assembly/council/committee which may have a number of associated sub-committees. The members of such a **planning authority** can be either government officials, political leaders, or a mix of both. The larger the proportion of political leaders in it, the more this authority is likely to be independent of the centre. The more leverage it will also have to impose its plan to the centre or to defend it at the stage of arbitration. For example, it is unlikely that a plan prepared and approved only by sub-national civil servants will stand up to central level, and will not generate frustration and disillusion amongst local planners.

As for the planning procedures, it is necessary to assess their **perenniality and the perenniality of the emanating institutions**. The more stable they are and unlikely to be dissolved or suppressed, the more independent they will be. Analysis should include whether the procedures are established by administrative circulars or rules, government orders, laws or acts of parliament, or under the constitution. The texts referring to the institutions' authority in planning should be scrutinized in a similar way.

ITEM	NO	YES	REMARKS
4. PEOPLE'S PARTICIPATION			
Local planning authority includes government officials only			
Local planning authority includes a mix of government officials and local leaders			
Local planning authority includes political leaders only			
Actions to promote people's participation			
Local planning authority legally established			
Its planning prerogatives are legally established			
People participate in term of labour/finance			
People participate in management of projects			
People are consulted during the planning phase: - by survey - by participation in assemblies and forums			

Government policy in the field of people's participation should be reviewed. In particular, actions taken by the government to promote people's participation in planning and implementation should be listed, to see whether government practice is in line with declared objectives. For example: provision of information to the population, formalities to be followed to create associations, encouragement given to associations and non-governmental organizations, flexibility for the use of locally mobilized resources (subject or not, to approval by superior levels), participation of leaders of groups or associations to the local planning authority, etc. This information will be part of the training programme where it would be used to make trainees fully knowledgeable about government policy. It will also give insights to trainers into how much support they can expect from the government for their training efforts.

People's participation at the implementation stage should be analysed. Do people only get involved in terms of providing labour, materials or money, or are they also taking part in or leading the management of local/village level projects ? This information will be of utmost importance while deciding on the content of an eventual training programme for the population or its leaders.

Lastly, **direct involvement of individuals** will have to be looked at to find out what information or training they may need, and also to develop an independent view on how best local plans can reflect the needs and demands of the majority of the population. This involvement can be direct (participation in a village assembly or forum where development projects or plans are discussed) or indirect (surveys, informal consultation of the population by its leaders).

Analysis of people's participation as above presented will provide trainers with sufficient information to make a first estimate of the amount of training to be given to non-officials, the size of the potential training target group and subjects to be covered.

2. TYPES OF DECENTRALIZATION

The framework to analyse decentralization which has been detailed in the preceding pages provides enough information to characterize the type of decentralization observed, along with some additional considerations on the stage of development of decentralization and the government commitment to it. The type of decentralization will be defined relatively to (i) administrative decentralization, (ii) financial decentralization, (iii) decentralization of planning and (iv) people's participation in development planning and implementation. The progressive pattern of the questions considered in each part of the framework makes the analysis easier. For the sake of easy visualization and characterization, a set of "marks" (between one and ten) have been assigned to the four aspects of decentralization and are at the basis of the patterns shown for the differents examples.

In **Nepal**, decentralization of administration has been implemented through actions for the creation of a local planning capability and a specialized planning cell. Specific decentralized planning procedures have been designed and made effective through cabinet decision. Local accountability of government staff is limited and some coordination committee exists at national level, although its real power is reduced. Administrative decentralization may be rated at 4.5 on a scale of ten 1/.

Financial decentralization is characterized by an extremely limited budget for decentralization, substantial allocation of funds to the districts but without a criteria-based formula, existence of a sizeable un-tied fund and of taxes raising power of local Panchayats and limited local financial accountability of government staff. Financial decentralization may be rated at 6.

Decentralization of planning is quite high in Nepal, according to the new rules, which in fact are not yet fully operational for many reasons. Demarcation of planning responsibilities has been made, although not very clearly; some training and technical assistance has been provided to local planners; location of district level projects is decided locally, and sectoral plans are prepared by the districts following the guidelines and financial ceilings provided by the centre. There is a local data base (Village and District profiles) and a Monitoring and Evaluation Committee is to

---

1/ See case study no. 4 on Nepal for more details.

monitor and evaluate the execution of the district plan. Intersectoral planning may take place within the various Plan Formulation committees and some integrated multisectoral planning could take place by using the un-tied panchayat funds. Projects are formulated locally and presented in unified formats, and the result of the planning process are village and district plans. Decentralization of planning may be rated at 7.5.

Decentralization in Nepal gives large emphasis to people's participation: the District Panchayat and the District Assembly are political bodies comprising elected members and leaders of associations. They are established under an Act and their planning responsibilities are stipulated by Decentralization Act and Rules. The population is called to participate directly in planning (at ward level) and in implementation (in terms of labour or finance and participation in project management through users's groups). People's participation can be rated at 8.5.

The analysis of the Nepal example helps to conclude that **decentralization in Nepal is people's participation oriented** (direct and indirect) **with substantial decentralization of planning.** Financial decentralization is less advanced and administrative decentralization is lagging behind. The stage of decentralization is advanced: institutions are set up, rules and regulations legalized. Schematically, this example can be represented by the following diagram:

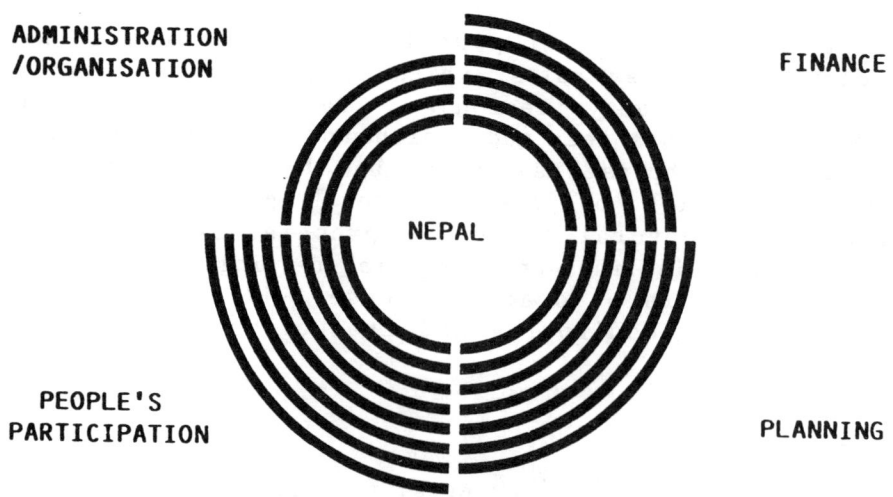

In this context, training activities developed in two complementary ways. On the one hand the Agricultural Projects Services Centre (APROSC) with the help of FAO implemented two related pilot training programmes: one on district agricultural planning and implementation and the other on decentralized planning. The first programme was addressed to government officials from the Ministry of Agriculture working at district and sub-district level and aimed at training them in programme, targets and project planning and implementation. The training approach included besides lectures, exercises, case studies and field work as well as post training on the job technical assistance. The second programme addressed both political leaders and government officials at district and village level. Five different programmes, one for each category of participants, were designed. The programme was developed after careful analysis of the field situation and also included post-training assistance for the preparation of a district plan. Training materials were elaborated and disseminated to the Panchayat training centres 1/ (Ministry of Panchayat and Local Development) through a trainers' training seminar, and a feedback seminar was held on the problems met during the implementation of decentralization. On the other hand the Ministry of Panchayat and Local Development and the National Planning Commission organized information seminars on decentralization, some special courses for local development and planning officers and the Panchayat Training Centres started a district level training programme.

In **Bangladesh** 2/, decentralization although very similar to that of Nepal has different characteristics. Administrative decentralization is much stronger, although the Government does not want to see a special planning cell established and thinks that officers involved in implementation at Upazila level should also do the planning. High calibre staff have been transferred to be the Executive Officers of the Upazila Parishads and there is substantial accountability of staff to the Parishads and staff assessment is conducted locally. There is also one central committee to coordinate and monitor the Decentralization Scheme. Financial autonomy appears also stronger in Bangladesh, although there are no un-tied funds as such. Resources have been earmarked to support decentralization and funds are allocated to the Upazila in a systematic criteria-based way. There is quite some flexibility in the funds which can be

---

1/ Regional training centres depending on the Ministry of Panchayat and Local Development, the mandate of which is to train political leaders and staff of the Ministry.

2/ See case study no. 2.

partially reallocated sectorwise. Moreover the final decision on financial allocation and the plan is taken by the Parishad and does not need to be approved centrally to become operational as is the case in Nepal. Decentralization of planning is slightly less than in Nepal: the lack of un-tied funds gives less opportunity for integrated multi-sectoral planning, there are no sub-national plans besides Upazila plans and no decentralized monitoring and evaluation units. People's participation is effective in the planning process through the Parishad, but direct participation is not institutionalized as in Nepal. Besides, people's involvement in implementation and management is much less.

Decentralization in Bangladesh appears to be more balanced than in Nepal. Compared to the other three axes, people's participation is slightly less developed, mainly because of the non-institutionalized direct participation of the population in planning and the lack of involvement in implementation and management. As in the case of Nepal, decentralization has reached an advanced stage of development.

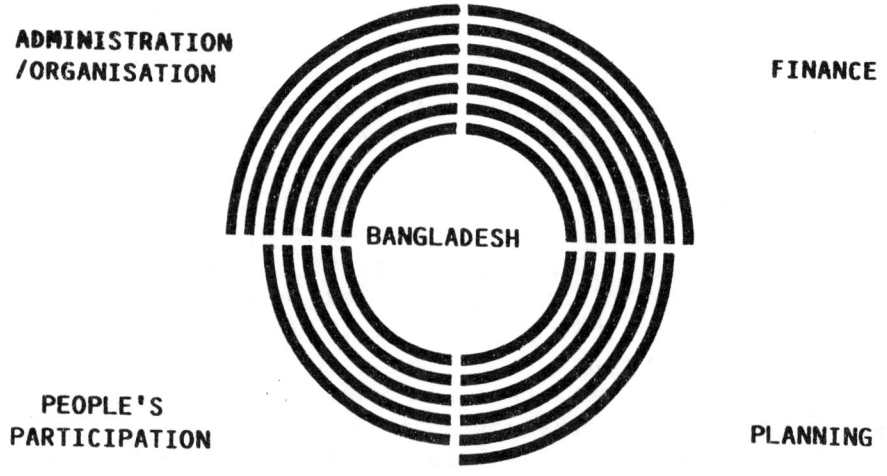

In this context, training/orientation programmes were organized by the two Bangladesh Academies for Rural Development (Comilla, Bogra) for Upazila officers and Upazila leaders. Following this activity, which was conducted separately for the two categories of participants, a more skill-oriented programme has been envisaged with the help of the Academy for Planning and Development which will involve a number of sectoral training institutions and will be coordinated by the National Training Council, Ministry of Establishment.

In **Tunisia**, decentralization started in the early 70's with the establishment of the Rural Development Programme under the office of governors 1/. This first effort was followed by the creation of a series of institutions to support decentralization: "les Offices" (parastatals in-charge of the establishment and management of large irrigation schemes which have seen their sphere of competence progressively expanded to non-irrigated and marginal areas) and more recently the COGEDRAT (Physical Planning and Rural Development Commission) and its regional 2/ and local (at governorate level) offices. These institutions have been given a planning mandate and have specialized planning units. The planning process in Tunisia was elaborated and fixed by official texts which are not implemented in reality. The COGEDRAT, chaired by the prime Minister, is responsible for decentralized planning but has not yet been able to impose its views to the line ministries. There were transfers of senior officers to the regions, particularly during the creation of the "offices". Responsibility of local officials is first towards the centre, but in the "offices", where the president director-generals are extremely powerful.

Financial decentralization is very limited. The only financial resources directly allocated to the governorates concern special programmes for rural development or emergency funds which represent only a small proportion of the total plan outlay. Decentralization of planning is also mainly linked to those special programmes which are to a large extent planned for locally: projects are formulated and some integrated multi-sectoral planning may take place like in the case of the PDRI (Integrated Rural Development Programme). National development objectives have been disaggregated in regional and governoratewise targets. There is a relatively elaborate production statistics system and some monitoring and evaluation is conducted by the local officers of COGEDRAT. There is virtually no people's participation in Tunisia. Political bodies like the Governorate Council have *de facto* very little to say in development planning and the population does not participate voluntarily in development works.

Decentralization in Tunisia appears to be mainly administrative through deconcentration or delegation ("offices"). Financial devolution and involvement of the population and its leaders are minimal.

1/   Tunisia is divided in Governorates headed by Governors nominated by the Central Government.

2/   There are six regions in Tunisia grouping each four to five Governorates.

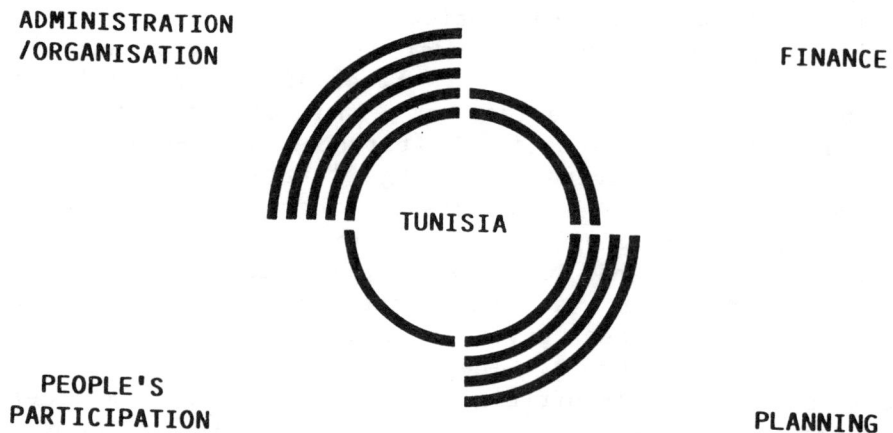

In this context, very little effort is being made to train for decentralized planning. Some training on project planning is organized by the "Centre National des Etudes Agricoles (CNEA)" where governorate level government officials may participate.

In **Niger** 1/, although the idea of decentralization goes back to the early 70's, decentralization is in a much less advanced stage of development than in Nepal or Bangladesh. With regard to the administrative and legal steps taken, the Government has taken action to create a local planning capability and a planning unit has been established in each "department". However the main planning work implemented at sub-national level will be done by technical officers from line ministries working at "arrondissement" level 2/ as the emphasis of decentralization is put on micro-projects planning. The planning process is being standardized but not yet in a clear way: many different approaches and approval procedures exist now. It is envisaged to transfer more senior officials to the "arrondissement" level. The administrative authority of the "prefet" and "sous-prefet" 3/ is remarkable (in terms of staff discipline, performance evaluation, finance).

1/   See case study no. 5.

2/   A "department" is divided into five or six "arrondissements".

3/   Officials in-charge respectively of a department and an arrondissement. The prefect is usually a strong political personality.

Financial decentralization is limited to the funds to be used for micro-projects and the government has not taken any formal action to encourage mobilization of local resources. There is no clear process of allocation of funds to sub-national levels: funds are in general granted for particular projects. They can be used for any kind of rural development activity, but the government discourages pure infrastructure and non-productive projects. "Arrondissements" and "Departments" can raise taxes. Local staff are financially responsible to the "prefet" and/or "sous-prefet".

Decentralized planning is de facto limited to micro-projects planning. The Government would also like to develop regional and physical planning, but it has not yet been clarified how. Some training on project planning and regional planning has been provided.

The decentralization policy in Niger gives a strong emphasis to people's participation through the newly created institutions of the "Société de développement". However it is a mixed technical commmittee comprising leaders and officials which is responsible for planning. The Government is promoting people's participation by training and information. Participation is both in terms of labour and involvement in decision-making during assemblies and forums.

Decentralization in Niger can be characterized as a partial decentralization insofar as it only focusses on one of the tools for development, namely micro-projects. Decentralization of planning is therefore very limited. People's participation and administrative decentralization are on their way but require to see rules, procedures and responsibilities clarified and legalized. Decentralization is still at an early stage.

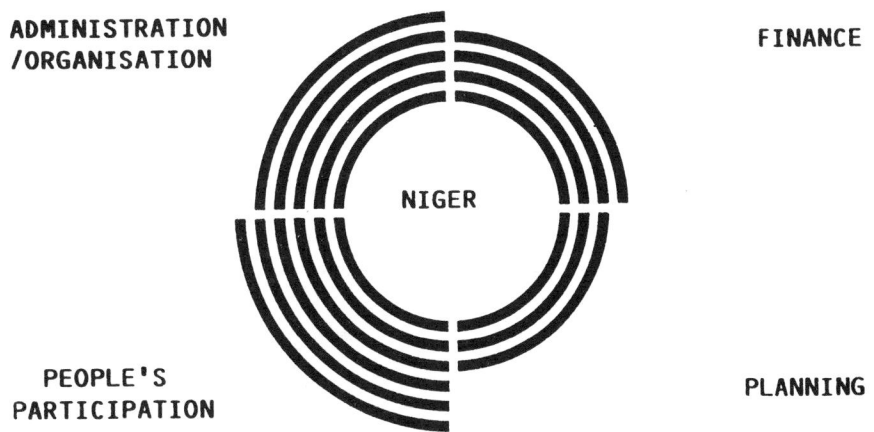

The training needs identified through an FAO project were thus restricted to the area of micro-projects and to the officers working at the level of "arrondissement" and "department". The programme elaborated had to start from scratch by first proposing a methodology to identify, prepare and analyse such rural development micro-projects. In addition to the training as such, technical assistance was envisaged to help the Ministry of Planning in establishing formats and procedures that would contribute to making this decentralization policy effective. Other bilateral and international assistance have been envisaged to enable the formulation and monitoring of micro-projects in a regional planning framework and to improve the financial management of the micro-projects. It is also recognised by the Government that concomitant training activities need to be developed with the people's organizations and cooperatives to enable them to become active interlocutors of the public administration. Viewing trained manpower needs at different levels and in direct relation to development planning and management has pushed the Government to consider the creation of a training cell at the Ministry of Planning that would guide and coordinate the different training activities envisaged.

3. IDENTIFICATION OF TRAINING NEEDS

The "framework to analyse decentralization" helps to identify what **type of decentralization** exists or is envisaged in a particular country and its **stage of development.** A certain number of statements can be made as a result of the analysis which are important for an eventual training programme:

- whether the government is committed to decentralization or not

- whether sub-national levels will be given resources and some latitude to plan for them

- what kind of planning has to be performed locally, and at which level

- whether the population is involved or not

- whether procedures are clear and can be integrated as a topic of the training programme or, if not, whether there need to be an experimental training-cum-action research programme.

The **main orientation of the type of training programme to be envisaged** can already be given at this stage: an experimental training-cum-action research programme, or simple training programme, training for officials only or for non-officials too, the size of the target, a content focussed more on technical matters (planning) or rather on behavioural and communication related topics.

Even a **preliminary assessment of the knowledge and ability** to implement a particular task and of the attitudes required can be made for the actors involved in decentralized planning (see Nepal and Niger case studies).

The analysis of decentralization as presented in previous pages will be the basis of a detailed identification of training needs. Detailed training needs identification, as the case studies presented show, is often not given the importance it deserves. Training programmes are frequently drawn up with haste once the decision to hold them has been taken. Past experience (7) points towards a certain number of tasks which should be executed in order to make sure that the training programme(s) envisaged will be adapted to the situation and be effective.

Decentralization by definition involves a **large number of actors** working at different levels. It is therefore necessary at first to list all those institutions involved in decentralized planning at national and sub-national levels, whether they are part of government administration or not. This task will give a general view of all those involved in the process for which training could be designed and organized.

Relationships and interaction among actors involved in decentralized planning and the general planning process in which they take place should also be analysed. Specific responsibilities can be delineated. This analysis could be summarized in a chart like the one presented on next page, where actors are in vertical direction, and time materialized by months on a horizontal axis. Planning functions are represented by boxes linked with arrows, like in a network. The chart shows the sequence of the planning functions as well as the flow of information or documents between the actors. It helps to identify key actors with most important planning responsibilities. The number of actors involved in decentralized planning being very large, it may be necessary, in view of limited resources available for training, to select those who play a prominent role and give them priority. This task will require extensive analysis of any documents published by the government on the decentralized planning process, discussion with policy and decision makers as well as field observation and discussion with sub-national actors in case the process is already implemented.

- 28 -

A detailed description of **functions to be implemented by the key actors** and the tasks they encompass should be made next. This description must necessarily become technical and is of the type found in any kind of training needs assessment. For example subjects like the content of planning documents or the format of the sub-national data base should be considered. In case decentralization is at a very early stage of development, it may be necessary to discuss with concerned organizations or institutions what kind of planning approach is to be adopted, given the manpower available and the proposed timeframe. Planning formats may have to be designed. The time available for performing a particular task would to a large extent determine the depth at which it can be executed, the assumption being that too big an increase of sub-national staff is not desirable, as it would create more recurrent costs and consequently lower the amount of funds available for investments. In such conditions, this task would become a real design/research activity on the planning process itself.

In case the decentralized planning process is already effective, this task would require a careful analysis of existing texts, an observation of what is really happening in the field, of the problems met, and an assessment of the necessary improvements to be brought in the present practice and eventually to the rules, in order to secure a successful implementation of decentralization.

Once the tasks of individuals are defined, the knowledge, skills and attitudes required to undertake such tasks would have to be identified (they will have already been outlined in the general analysis of decentralization). **Knowledge** will in most cases include planning rules and procedures, the objectives of decentralized planning, the development objectives of the country, planning concepts, and understanding of what other actors of the planning process are doing. **Skills** will comprise the command of particular tools and techniques necessary to implement given tasks. **Attitudes** will be most often associated with behaviour required to assume the new responsibilities given to the planning actors, for example: for local leaders, to be more development oriented; for local government officials, to be more open-minded towards local leaders, be more cooperative with other officials, have more initiative, etc.; for central level officials, to be more open to local level suggestions.

Next a **comparison between present planning practice and desired practice under decentralization** could be conducted. Actual planning practice has to be analysed in depth by observation, review of planning documents produced, interview with the actors of planning and their supervisors. Presently executed tasks should be documented as much in detail as future tasks. For each key actor a comparison between present performance and future expected performance should be made and differences identified. These differences should be linked with the knowledge, skills and attitudes described previously, and, as a result, training objectives should be derived.

Following this qualitative analysis, a quantitative approach to the problem could consist in the analysis of **staffing patterns** in the offices involved in decentralized planning. The number of staff of various levels and designations can be determined. Although, it is assumed that as few as possible new staff should be recruited, estimates of additional requirements could be made, keeping in view the minimum quality required for planning work and the time necessary for it. This could be made by comparing present staffing patterns with expected future planning responsibilities under decentralization.

In the finalization of the assessment, two types of training needs may emerge: in-service and pre-service. Pre-service training needs correspond to the training to be provided to the additional staff to be recruited. We will not give much attention to this type of training, as it is expected to concern only a small number of people in the short run. In-service training needs, on the contrary, represent the training to be provided to the actors who are already at duty and simply need complementary training to make them able to cope with their new responsibilities. For each type of actor the following information should be given: designation; number to be trained and knowledge, skills and attitudes to be developed through training. Besides, suggestions on areas in which special attention is needed (in terms of post-training technical assistance for example) should be made at this stage.

This process, summarized in schematic form on the next page, can be followed in different ways, but requires anyway **a good knowledge of the texts defining the planning procedures and extensive interaction with both potential trainees and the heads of client institutions.** It can be a joint exercise between the training institution and users as in the case of BARD Comilla in Bangladesh (see case study no. 2). It can also be an on-going process like in the case of RDTRI, Colombo, in Sri Lanka (see case study no. 6) where monthly or quartely sessions are organized with trainees to collect their feed-back and suggestions, or like in the case of the FAO/APROSC Training Project in Nepal were the training needs are first identified through observation of trainees at work and their assessment modified year after year in view of their post-training behaviour (see case study No. 4). Special committees, seminars, surveys, courses, on-the-job visits can be used for the purpose.

Identification of training needs
Summary chart

STEP 1	General Analysis of Decentralization (Framework)
STEP 2	List all the actors involved in decentralized planning
STEP 3	Analyse interactions between the actors
STEP 4	Description of tasks to be implemented by <u>key</u> actors
STEP 5	Identify knowledge, skills and attitudes required to implement these tasks as well as need in manpower
STEP 6	Compare actual practice with future practice
STEP 7	Analyse the staffing pattern
STEP 8	Finalize training needs assessment

PART II. TRAINING FOR DECENTRALIZED PLANNING: SELECTED ISSUES

The training programmes reviewed for the preparation of this paper, some of which are described in more detail in Volume II (Case Studies), were selected to reflect **diversity both in terms of the type of decentralization envisaged and of training approach used.** They represent only a fraction of the experiences conducted throughout the world, but they are representative of the total. They include examples of countries involved in decentralization over a long period (India, Nepal) and others which have come to it more recently (Bangladesh, Niger). Some give more weight to the technical/administrative aspect of decentralization, others to the political change it brings forth. The examples also illustrate different stages of development of decentralization.

The cases selected also illustrate a large gamut of the training approaches followed in relation to the type of decentralization. The concept of training approach encompasses here the combination of a number of characteristics relating to the objectives, the organization and even to a large extent the content of the training programmes. These characteristics refer to their level (international, national, regional), their participants, their development objectives (individual, institutional), their stage of evolution (pilot, nation-wide) and the training methods used (with reference to duration, frequency and timing of contacts, teaching techniques, post-training activities, etc.). In some of the cases presented, the relation between the type of decentralization and the training approaches was explicited at the design stage (e.g. Nepal, Niger), in others the link was less direct or restricted to some aspects of the decentralization envisaged.

The analysis of individual cases of training for decentralized planning conducted during the preparation of this paper has mainly consisted in replacing the training programmes reviewed in the context in which they developed, this context being understood by using the framework presented in Part I, and analysing the linkages between training and decentralization. The results of this analytical work will be presented and organized according to the major issues identified.

1. TIMING OF THE TRAINING PROGRAMME

Decentralization appears to be going through a **certain number of phases of development** which are summarized briefly here. The very first phase is the emergence of the will to decentralize which can be professed in speeches of political leaders or proclaimed in official documents like a constitution (Nepal). In the second phase this will often crystallize in the establishment of structures which will be the basis of decentralization. These structures can be techno-administrative (Regional Development Authorities - Brazil, Philippines - ; Regions with their planning units - Zambia, Tunisia - ; Development Agencies - Morocco, Tunisia) or politico-administrative (Development councils - Bangladesh, Zimbabwe, Niger, Nepal). The third phase which is sometimes cut off is the publication of a text fixing the general principles that are to rule decentralization (Decentralization Act 1982 in Nepal). The fourth phase is the publication of detailed rules for all agencies or institutions involved (Bangladesh 1983, Nepal 1984, Karnataka (India) 1986, Zambia 1980). The fifth phase is the operationalization of the rules (see chart next page).

A phased development of decentralization

According to the stage of decentralization, the type of training activities undertaken are quite different. That is the reason why the framework to analyse decentralization insists so much on this aspect. Experience even shows that **there seems to be a stage before which it is not advised to start organizing training.** Two examples may illustrate the point in question.

When to undertake training

In 1980, following a request of the Government, FAO undertook a training programme for decentralized planning and implementation in **Niger.** The will to decentralize had been declared, but it had not yet materialized in terms of structures and rules. The evaluation of the programme conducted late 1982 showed that the impact of the training was limited, in particular because the trainees, who were government officials working at sub-national level, did not yet have the reponsibility to plan. Furthermore, the programme did not get all the necessary support from the central government as decentralization had progressively lost its importance. It failed and probably even created some frustration among the participants who were taught skills they were not allowed to use.

Examples of programmes anticipating on decentralization

Niger

Nepal

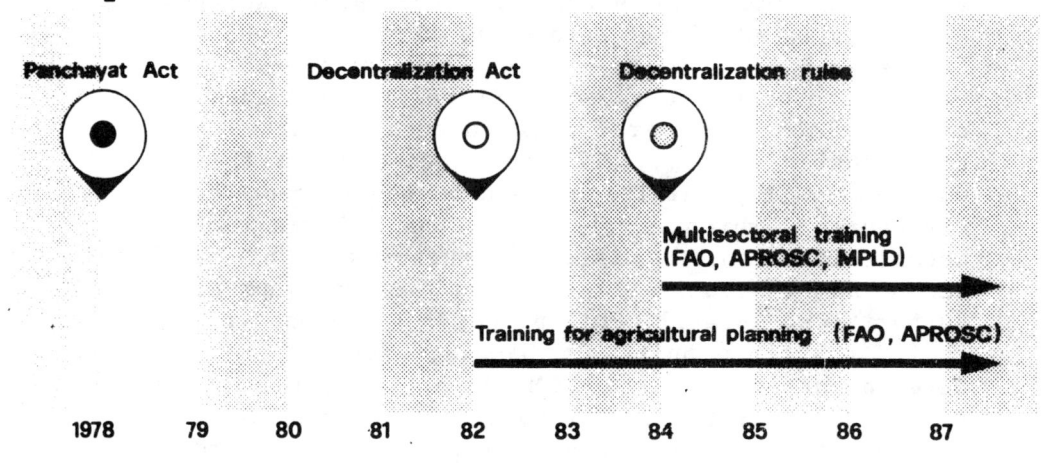

- **Panchayat Act** (1978) ●
- **Decentralization Act** (1982) ○
- **Decentralization rules** (1984) ○

Multisectoral training (FAO, APROSC, MPLD) → (1984–87)
Training for agricultural planning (FAO, APROSC) → (1983–87)

1978 79 80 81 82 83 84 85 86 87

Niger

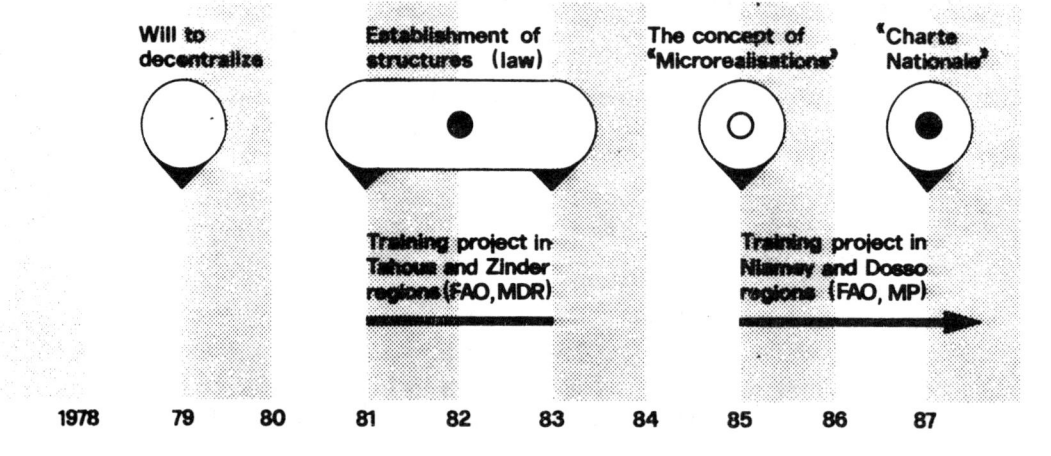

- **Will to decentralize** (1978)
- **Establishment of structures (law)** (1981–83) ●
- **The concept of "Microrealisations"** (1985) ○
- **"Charte Nationale"** (1987) ●

Training project in Tahoua and Zinder regions (FAO, MDR) (1981–84)
Training project in Niamey and Dosso regions (FAO, MP) → (1985–87)

1978 79 80 81 82 83 84 85 86 87

Bangladesh

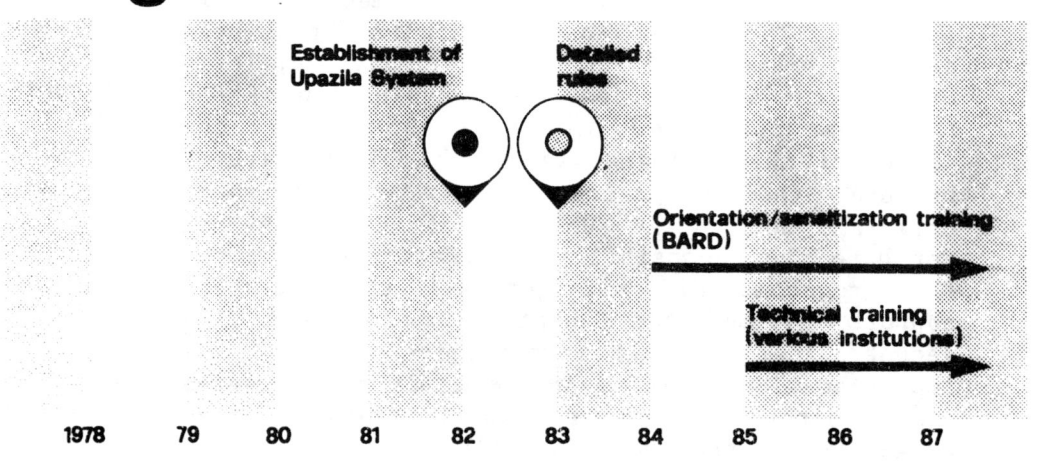

- **Establishment of Upazila System** (1982) ●
- **Detailed rules** (1983) ○

Orientation/sensitization training (BARD) → (1984–87)
Technical training (various institutions) → (1985–87)

1978 79 80 81 82 83 84 85 86 87

Karnataka

The Institute of Social and Economic Change (ISEC) in Bangalore, Karnataka in **India**, launched a training programme for district planning in 1973/74. The programme consisted in training the District Planning Officer (newly established), the District Agricultural Officer and one College Professor in nineteen districts. Following the course, each team had to prepare the perspective development plan of the district where they were posted. In fact, many participants got transferred and teams were disbanded. In the cases where plans were prepared, inadequate resources were available to implement them: sectoral allocations decided at state and union level did not match with the requirements of the plans. As a result the programme was abandoned. Clearly, the programme was too much anticipating on decentralization and the conditions (particularly the financial procedures) were not yet adequate to allow participants to apply what they had learned. These examples do not preclude that training could in theory prepare the ground for the envisaged decentralization. Purportedly successful experiences of "anticipatory training" are not known.

Concomitant development of decentralization and training

Experience shows that **training can be most effective and useful when decentralized institutions have been established and when the government is trying to define their working procedures.** In this case, the training should be more of an **action-research type** and of a limited coverage (one pilot region for example). Training can help to test rules and procedures proposed by the government. It should be accompanied by technical assistance and the programme should have a strong built-in monitoring mechanism which can provide feedback and suggestions to the government on how the rules and procedures can be improved. The case of **Nepal** illustrates somehow this kind of situation: problems and constraints experienced at the district level during the first two decentralized planning cycles were fed-back to policy makers through seminars or meetings. Suggestions were made on procedures, planning formats and demonstration was given that certain rules were impossible to be followed by the districts because they were too demanding in manpower.

Nepal

Niger

In the case of **Niger**, the training programme elaborated for the second project was in a position to make a proposal on the format to be used for presenting micro-projects and was also requested by the government to make suggestions on how to monitor and evaluate micro-projects and on how micro-projects could be planned in a regional development context. Such contributions which can be of crucial importance at the stage where decentralization has to be made operational, can only develop if the training programme has strong links with the ministry/institution in charge of decentralization. Such link can be materialized for example by a programme steering committee where this ministry/institution is represented at a sufficiently high level or by locating the programme in the ministry/institution itself.

Phased programme

Bangladesh

As for decentralization, **training itself may be organized in phases as well.** The example of **Bangladesh** is very illustrative in this context. Following a careful observation of the planning process at upazila level, an orientation programme geared towards a vast and diversified audience was designed and implemented by BARD Comilla and BARD Bogra. Subsequently, it was felt that although this first programme had been instrumental in informing and sensitizing a large public, it had not solved the problems faced during the plan formulation. The need for more technical training on planning was acknowledged and a larger programme is being prepared now which will involve a number of sectoral technical training academies.

These few examples illustrate the importance of the timing of training activities. The framework to analyse decentralization includes questions which help to assess the stage of development reached by decentralization as well as the government commitment to it. These two indications will be capital to decide (a) whether the situation is ready for training and (b) for what kind of training (action-research, pilot, orientation, technical, massive, etc.).

2. PARTICIPANTS IN THE TRAINING PROGRAMME

By nature, **decentralization affects a large number of people by modifying their responsibilities.** According to the type of decentralization, in different countries, the number and kind of people concerned will change, so will the alteration in their duties. A systematic identification of training needs will provide a clear-cut answer to this problem. However in the process it will be necessary to select the key actors that will have to be trained. This selection will have to be done carefully as, if the choice is inadequate, the impact of the training programme is likely to be relatively small.

Who should be trained in priority

The example of the **Nepal** training programme is very explicit. The analysis of the decentralized planning process and the observation of a planning cycle in a district had demonstrated to trainers that ward chairmen, ward members and more generally the whole population had to play a fundamental role and needed to be informed and oriented, in order to come up with good project proposals concerning all aspects of development for inclusion in the plan. Various constraints, in resources as well as in competence, led trainers to abandon this component of the programme. The consequence was obvious: planning at village and ward level did not improve as was expected and it had to be acknowledged that training only village leaders and the village Panchayat Secretaries was not sufficient. Careful identification of training needs based on an in-depth analysis of decentralization with the use of the framework prevents from selecting inadequate clientele groups for training.

Population

The type of decentralization will largely determine the participants in the training programme. In **Kerala** (India) for example where decentralized planning is limited to the preparation of a Special Component Plan for scheduled castes and tribes, it is the district officials who do all the planning with some advice from local political leaders. Training will therefore be concentrated on the former. In **Zambia** and **Niger** where local planning and identification and formulation of development

District level officials

activities are by and large the responsibility of government officials, it seems justified at present to start training the latter first, as resources, both human and financial, are very limited. In **Bangladesh** and **Nepal**, the local political bodies play to the contrary a leading role in the planning process as they are initiating it and have its overall responsibility: this implies that their members must be trained.

Local political leaders

As decentralization implies a change of relationship between the centre and the regions, central level officials from Departments and Ministries will also have to be concerned. They are the actors who are usually reluctant to decentralize and who do not trust local leaders and officials, and training will have to induce a change in their behaviour.

Central level officials

3. THE CONTENT OF THE TRAINING PROGRAMME

Decentralization implies changes in planning and implementation procedures and in responsibilities of all those involved in development activities. **All actors concerned should therefore first of all be informed of the new procedures and rules,** of the reasons and objectives of the change occuring, and of what is expected from each of them in particular. It is not uncommon in countries where decentralization is being carried out to observe that council members or officials do not know that they have to do: meetings are being organized and directed by the central government but no one exactly knows what should be discussed in them. In more extreme cases still, individuals may be aware of being a member of a particular committee but without actually knowing what the committee is for. In such circumstances it is evident that decentralization cannot succeed and its detractors will find it easy to criticize it. Necessarily, information on decentralization will have to be provided to the largest audience. Resources will be needed to publish books and pamphlets, prepare posters for illiterates and distribute them and organize radio broadcasts, seminars and meetings.

Information on decentralization

Decentralization gives new responsibilities in fields that people may not be conversant with or may not even be aware of. Political leaders used to political debates or to lobbying in order to get infrastructures constructed in their constituency suddenly have to adopt the budget for their region, decide on taxes and select projects. Government officials accustomed to implement programmes imposed on them by the central government, all of a sudden have to assess the development potential of their region, define priorities, design programmes and formulate projects; technical officers have to get involved in planning and administrative work. Such drastic change in activities may require very basic training where first of all the major concepts are introduced and then progressively tools and techniques are provided. Subnational staff and leaders will have to be briefed on national development policies and objectives as well, so that their proposals be in line with them and consistency between national and decentralized plans be secured.

Decentralization also entails changes that call for radical modifications in behaviour. In a centralized system, government officials working at subnational level have to execute orders given by the central department to which they belong; coordination between field offices is weak and they have a tendency to work independently. Overall coordination is supposed to take place at a higher level. In a decentralized system, the picture is reverse. Government officials have to take initiatives and translate broad national policies and objectives in a plan adapted to the specific conditions of the area where they work. Central departments from "dictators" become the points were eventual bargaining, coordination and discussions take place. Coordination between field offices must take place at subnational level both for planning and implementation.

Inducing behavioural changes

In order to facilitate this fundamental behavioural change, training must provide specific inputs (a) to **give confidence** to the subnational staff that they can take upon themselves their new duties, (b) to **assist the staff in getting organized** as a coordinated planning and implementation group and convince them that organized as a group they will be more effective; (c) to **make them aware of intersectoral linkages** and provide them with some simple tools to deal with them, (d) to **make them more receptive** to, and **cooperative** with, the local political leaders and the population as a whole as the latter is likely to be increasingly involved in development related decisions under decentralization, (e) to **modify central department staff's opinion** on what subnational staff can achieve on their own initiative in the field of planning. The relative importance of these inputs will certainly vary according to the specific characteristics of decentralization.

The training programmes reviewed very largely neglected the issue of behavioural change of government officials. Some attempts to deal with it, but only partially, can be observed in **Sri Lanka, Niger** and **Nepal**. Giving confidence to subnational staff and assisting them to get organized as a group could be achieved through group dynamics sessions where members of actual working groups would interact. By having to achieve collectively tasks of a growing complexity, groups would automatically get organized, leadership emerge and confidence grow.

Intersectoral linkages

Intersectoral linkages can be best grasped through examples or case studies where their oversight resulted in failure of programmes and/or unexpected problems. Tools to deal with them could for example be impact analysis where all major consequences of a particular activity are considered upstream as well as downstream, and complementary action is identified in concerned sectors. This topic is well coped with group simulation as tried in a simple way in Nepal.

Mutual knowledge and understanding between the population and government officials

In many countries, it is common to find **reciprocal negative attitudes of the public administration and the population.** Such state is certainly not favourable to decentralization if devolution is an important aspect of it. One possibility of improving the situation is to improve the mutual knowledge and understanding of these categories. This can be done directly and/or indirectly. Indirectly, it supposes to explain to one group the behaviour of the other in terms of objectives, constraints, etc.; directly, it implies to organize interaction which presents the risk, if it fails, to further degrade the situation.

In most cases, **political leaders under decentralization are anticipated to be increasingly involved in development management.** They are expected to give paramount importance to development issues affecting their community or administrative unit and to perceive development management no longer as a matter of using resources given by the central government, but as a process of managing internal and external resources in order to achieve well defined development objectives reflecting the potential and needs of their region. They are also to become more and more conscious of the need to design feasible projects and plans.

<small>Sensitization of political leaders</small>

In the process they are to develop cooperative working relationships and understanding with local government officials. For this purpose, it is necessary to make sure that they be familiar with the concepts used and the mechanisms referred to by the latter.

Experience in this area is rather limited and adapted approaches would need to be designed and tested. The same applies to favouring interaction between leaders and government officials. Whilst in theory it would appear as both desirable and necessary, in practice it is widely avoided mostly because it is a difficult and risky area where little known conclusive experience has been accumulated.

<small>Interaction between leaders and government officials</small>

Transformations induced by decentralization will naturally necessitate acquisition by actors concerned of new knowledge and skills of technical nature. Although varying in importance according to the type of decentralization and specific country situation, this content can schematically be organized in four parts: (a) situation analysis, (b) project planning, (c) plan formulation and, (d) implementation.

<small>New knowledge and skills of technical nature</small>

Data collection techniques (questionnaire and survey design) have been assigned considerable time in most training programmes surveyed. Techniques presented can lead to the establishment of elaborate data bases or profiles as in the case

<small>Data collection</small>

of Karnataka, Kerala, Nepal or to quicker and more qualitative data for rapid appraisal of the situation like in Niger. The choice of the methodology proposed depends essentially on the institutional set-up existing at sub-national level: in case a special statistical and planning unit exists, it is possible to collect more data. Trainers must however always be sure that data collected will be useful and used and that resources exist both for collection <u>and</u> processing and analysis.

Data processing and analysis

The **processing, analysis and use of elaborated data** are not always dealt with much detail in the programmes reviewed. The perspective in which they are treated also varies according to which aspect of planning is emphasized. It may be rapid rural appraisal for project formulation (Zambia, Niger), spatial analysis (IPD/AOS, UNCRD 1/, Sri Lanka), problem, need and potential identification (Sri Lanka, Nepal), regional analysis (UNCRD, IPD/AOS), etc. The diversity observed in this part is symptomatic of the variety of participants in the courses, their various objectives and the availability of data. It is nevertheless possible to say that a training programme for decentralized planning should include inputs on data collection (both first and second hand) and analysis in a regional perspective (with spatial considerations). Even if only a limited part of planning is decentralized, like in the case of Niger where it only concerns the formulation of micro-projects, the need to define a regional framework in which planning can take place, arises rapidly.

Demographic analyses

The virtual absence of consideration of population issues in the training programmes reviewed is an illustration of a long lasting chronic deficiency in development planning. This lacuna results in a static description of the situation of a given area, the oversight of essential matters like employment, migrations, changes in man/land ratio and the inability to forecast important problems which necessitate early action.

1/ IPD/AOS: Institut Panafricain pour le Développement.

UNCRD: United Nations Centre for Regional Development.

Project planning is generally taught in a conventional way. Concepts and techniques presented are usually those found in courses for large investment project planning : Net Present Value, Benefit Cost Ratio, Internal Rate of Return, etc. In some cases, the programme even includes economic analysis (IPD/AOS). This practice is not satisfactory for two main reasons. First, **it does not seem possible to use the same techniques for the small and medium sized projects formulated at decentralized level as for large investment projects**: it is not possible to spend considerable human and financial resources to plan for a small investment. Investigations specific to a particular project have to be limited, project formulation must be completed in a relatively short period and the productivity of project planners in terms of number of projects prepared must increase. If not, the benefit expected to accrue because of the use of the project approach will be outbalanced by the costs incurred for putting it in practice. Second, projects formulated at decentralized level are in a large proportion social sector projects, the selection of which will not be based on profitability criteria.

Up to day, there is no widely recognized solution to this issue and the field is open to research and experiment. One way of going about this problem may be to **combine simplified project planning with regional analysis.** Simple technical descriptions and standard costs of the most common projects can be prepared (Kenya (8), Kerala, Nepal, ...) at the subnational level with eventual help from the central departments, and be updated regularly. The results of a regional analysis will provide on one hand a guide on what kind of projects could be undertaken in a particular area and with what degree of priority, and on the other a set of simple criteria which can be used to select projects and which will have to be analysed for each proposal. This method would allow for people's participation at the stages of project identification and selection, and should of course leave the possibility to take liberty with the standard projects and adapt them to specific conditions. As the projects presented would a priori have been ascertained to be both technically and financially feasible,

Project planning

decentralized planners could concentrate more on issues related to institutional set-up for implementation of the project, manpower requirements, population participation, etc.

In terms of training, such an approach would imply intensive and advanced training in project planning and regional analysis for a few (the specialists), familiarization with the standard projects and the results of regional analysis, presentation of project formats and training on techniques related to implementation planning and people's participation to the mass of development workers.

Plan formulation

Among the training programmes reviewed, only a few provide inputs on **plan formulation**. The reason is that in many countries decentralized planning is limited to project planning and decentralized plans if they exist are little more than an aggregation of projects. This fact notwithstanding, a certain number of inputs should be provided by a training programme : project selection techniques, intraregional balance analysis of the plan, review of intra and intersectoral linkages of proposed projects, their implication in terms of support activities and manpower, etc. The importance to be given to this aspect of the content will largely depend on the kind of planning advocated under decentralization and specially the extent of emphasis given to intersectoral/multisectoral integration.

Management topics

Decentralization multiplies responsibilities of subnational staff in the field of implementation, monitoring and evaluation. In Bangladesh for example, specific training inputs were provided to the participants to the programme on financial management and accounting. The same is the case in Karnataka (India) and Zimbabwe for local government staff and in Ghana (Economic and Rural Development Management (ERDM) Programme). With the growth of local programmes likely to take place under decentralization, the need is also felt to improve office organization and time planning of activities. Training in these matters was provided for example in Sri Lanka and Nepal (networking, scheduling, task analysis, information analysis). In the field of monitoring and evaluation, special courses or modules have been developed in Zambia, Sri Lanka and Nepal to train subnational officers on the design and operation of simple monitoring systems (logical framework).

The number of different participants concerned and the diversity of topics covered and objectives pursued, clearly indicate that **training for decentralized planning cannot be a simple venture.** It will generally be a programme comprising a series of courses drawing on a wide range of expertise and using a gamut of organization and management approaches.

4. INSTITUTIONAL SET-UP FOR TRAINING

The specificities of decentralization and of the related training define broadly the institutional set-up in which training programmes should be organized and the way they should ideally develop. Decentralization is first a political venture which can be envisaged for a variety of reasons: development of democracy at the intermediate or lower level, establishment or revival of subnational political bodies, distribution of central power to a larger number of decision making units, management of inter-regional balance, etc.

As a political undertaking, decentralization often emerges suddenly to become an urgent priority. This implies it has to be implemented rapidly if not in haste. It is also a technical endeavour aiming at improving the design and implementation of development programmes and increasing popular participation in development in terms of resources as well as decision making. As a stated principle, at first, decentralization has to be translated into a process and rules, and necessitates in most case the elaboration of laws – a time consuming procedure –, and the execution of a vast training and information programme. This sudden rush suggests that the planning process and rules are likely to change after actual implementation has started. It confirms the **potential role of training as a feedback and monitoring tool** and the absolute necessity of a

Rapid implementation of decentralization

Links with training

strong link between whoever trains or informs and the government unit in charge of the implementation of decentralization: trainers have to be perfectly aware of the latest decisions taken, of the latest rules and/or formats adopted.

As a sudden reorganization, decentralization necessitates an unforeseen and often huge training effort which will relatively rapidly decline. Existing government officials and subnational political leaders have to be trained at short notice: the **training demand thus increases far beyond the capacity of existing training institutions** in quantity as well as in quality as **it encompasses subjects which usually are not covered under existing programmes.** In Nepal for example, a crash programme designed for the Panchayat Training Centres 1/ was intended to involve yearly fifteen districts out of a total of seventy five in short training programmes for district officers and District Panchayat members. A more in-depth training programme, similar to what is described in the Nepel case study, would require to more than double the existing staff of the centres and a substantial increase of their budget. Once this first wave is over, the training demand would be limited to refresher courses and training for new staff and newly elected leaders: this reduced need would not justify the additional staff established earlier.

Need for increased training capacity and in new subjects

A large proportion of the objectives of training programmes for decentralized planning relate to skills to be developed among the participants. This implies that the training must be practical and action oriented. **Trainers involved need to have a good field experience** and knowledge as well as tools and techniques to be taught have to be carefully selected and adapted to fit to a specific context. Unfortunately the staff of existing specialized training institutions are in most cases involved only in teaching and academic work and lack this kind of ability. Trainers also need to have an excellent ability to assess the situation, make suggestions to the government and have a high degree of motivation and mobility.

1/ Regional training centres belonging to the Ministry of Panchayat and Local Development the mandate of which is to train political leaders and staff of the Ministry.

Three examples of response to these specific conditions will be presented shortly herebelow. In **Bangladesh**, all existing training institutions will be mobilized to face the demand for training on decentralization. After a first phase devoted to the information of political leaders and district level government staff by the two Bangladesh Academies for Rural Development (Comilla and Bogra), the Government has decided to organize skill development training programmes through existing sectoral academies. These programmes will be coordinated by the national Training Council, Ministry of Establishment which coordinates all training activities in the country and approves the yearly programmes proposed by the training institutions. As a preparation, the council will organize a trainers' training on local level planning with the Academy for Planning and Development. Subsequently, trained trainers from other academies will design sectoral training modules and handbooks, and courses will be organized for a wide public. In this scenario, preference is given to the **mobilization of maximum existing training resources**. The link between the programme and decentralization is secured through the Council.

A number of training institutes

In **Nepal**, the training programme in its pilot phase developed in the training wing of a parastatal consultancy firm (the Agricultural Projects Services Centre) comprising a limited team of trainers with practical experience in project planning and previous experience in training for district agriculture planning. Overall guidance to the programme was provided by a high level Steering Committee of which major policy makers involved in decentralization were members. During the initial phase, feedback on decentralization was channelled from the field to central government through the committee and a national seminar, and training materials were developed. Progressively as experience was gained, it was felt that trainers' training programmes should be organized. A first programme took place to train instructors from the Panchayat Training Centres and more are planned for the future. In this case, care was given to have **the training programme developed by trainers having some practical field experience through an action-research kind of programme, and in direct relation with policy makers**. Massive training is expected to take place through existing training institutions.

Training wing of parastatal consultancy firm

Ministry of Planning

In **Niger**, the training programme developed within the Ministry of Planning, in the division in charge of micro-projects. The training methodology and materials were elaborated with young professionals working in the division. The programme produced proposals for planning formats and it is expected to make proposals on the programming and monitoring of micro-projects at the regional level. The development of a national training capacity in this field has also been initiated: ten staff of the Ministry of Planning have been trained as trainers and will be involved in preparation of micro-projects in the field. It is now proposed that these officers besides their usual duties will be made available for training programmes as needed: this duty will be added to their existing terms of reference. Contacts have been taken with the Institut Pratique de Développement Rural of Kolo, that trains most rural development workers in Niger, in the hope of their becoming involved in the programme in the near future. In this example, **maximum emphasis has been given to link training with decision making.**

5. THE NEED FOR A COUNTRY SPECIFIC APPROACH

Our analysis of training for decentralized planning has continuously pointed towards the need of a country specific programme: **decentralization itself develops in a unique way in a particular country according to its political and institutional set-up and the training must be designed to fit to it.** The three cases of training developed outside an explicit politico-institutional framework demonstrate a contrario this assertion.

India

The programme run by **AVARD** 1/ in India was designed to meet the need of a number of voluntary agencies working in the country and facing different political and institutional situations. Decentralization, although strongly encouraged and monitored by the Government of India, is the responsibility of State Governments that are free

---

1/ Association of Voluntary Agencies for Rural Development, Delhi.

to decide in which way (procedures, institutions, etc.) they are going to decentralize and at what pace. Situations vary much, ranging from very little (Punjab) to substantial (Maharashtra) decentralization. Consequently, AVARD developed a technical training package for local level planning and concentrated on training only staff from its member agencies. As a result, plans formulated were in their great majority not implemented and concerned government officials are generally unaware of their existence.

The course organized by the **Panafrican Institute for Development/West Africa (IPD/AOS)** in Ouagadougou, Burkina Faso, is an example of post-graduate training in regional development planning. Its objective is to train regional planners (generalists) able to coordinate planning at subnational level. Governments have used trained participants to staff their sub-national planning units and have confirmed their operationality (Burkina, Niger, etc.) but requests have come recently (Niger, Guinea, Senegal) to have country programmes for a larger number of officers.

West Africa

The **United Nations Centre for Rural Development (UNCRD)**, Nagoya, Japan, has been involved in training for regional planning since 1971. During the first years, the course was of four to six months duration, highly technical and included practical fieldwork in a selected country (see case study No 9). One major lesson learned from these courses was that training benefited more trainees from the country where the fieldwork was organized than others. Consequently, UNCRD reduced the duration of its general course to two months and started to get involved in country programmes (Sri Lanka).

UNCRD Nagoya

A rapid review of programmes run at regional/international level supports the view of their usefulness for general educational purposes and for promoting a regional approach to development. At the same time it shows their limitations when viewed against specific country requirements. The relevance and applicability of general regional planning training for developing country officials have been dealt with extensively elsewhere (9).

Role of international training centres

International training centres, nevertheless, appear to have major roles to play. They can, in fact, provide **specific technical training** for small numbers of specialized planners (regional accounting, statistics, regional analysis, etc.) and, more generally, they can **backstop and assist national institutions** involved in training for decentralized planning. They are also ideally suited to be a **forum** where policy makers and planners can exchange their experiences on decentralization.

SUMMARY AND CONCLUSIONS

The implementation of a decentralization policy is a gradual process where the changes required in the knowledge, skills and attitudes of the actors involved play a vital role. The training programmes analysed have all attempted to elicit such changes to varying degrees and they have been viewed as a means to contribute to the overall process. It may be expected that demand for such training programmes will increase in the near future in view of the interest shown by a larger number of developing countries for decentralized planning and implementation and the pressure by the international and bilateral technical and financing agencies on fostering locally-formulated self-supporting development activities.

Framework to analyse decentralization

This text has been elaborated with the objective of disseminating information on a number of different decentralization experiences and related training programmes, drawing attention to some selected issues. The cases presented have been reviewed by using a **framework** that aims to be an analytical tool enabling trainers, faced with the task of designing training, to acquire full knowledge and understanding of the decentralization process.

The framework is structured so as to give information on the type of decentralization being implemented or envisaged, its stage of development, the extent of government commitment and the responsibilities of the actors involved. Trainers are thus able to assess the degree of administrative deconcentration and delegation, of financial devolution, the degree of complexity and deconcentration of the planning process and the importance given to people's participation. The knowledge of this context and of the tasks to be performed will help determine the objectives to be set to a training programme, its content, organization and audience.

This proposed way of looking at training for decentralized planning leads almost inevitably to **nationally conceived programmes best implemented by national institutions.** These can assure a continuous link with the type and pace of decentralization through their constant contact with the government unit(s) responsible for decentralization and with the practice of planning and implementation through visits and contacts at the local level. National institutions can best organize and have, as regular part of their programmes, post-training support activities (technical assistance), a usually neglected area. It is through such activities that valuable feedback can be provided to the process of decentralization and to adjustements required in the content of training to better reflect the practice of planning. The management of such training programme will imply a continuous assessment of the evolution in job performance of trainees as well as close monitoring of the changes in the environment in which they work. It also assumes that the links between the training institute and the government units responsible for decentralization are set up and formalised.

Need for national training programmes

Institutional and operational link of training with planning units

As regards the **contents**, experience shows that training requires to be designed for both **instilling knowledge, teaching applicable skills** as well as **inducing behavioural changes.**

"Administrators" are expected to become development oriented officials and as such to develop initiative, favourable attitudes with regard to the population as well as abilities to

Training content

plan in a multi-sectoral and spatial framework that secures complementarity and coordination of development activities. Such improvements in behaviour and attitudes can be brought about also through training, although not much has been attempted up to now.

The **type of knowledge and applicable skills** for planning to be taught through in-service training is also a debated issue. A general demand is there for adapted methods and tools in areas like linkages between national-regional-district level planning, demographic analyses, planning of small-scale projects, their selection and prioritization in a regional planning-programming framework, intersectoral integration of projects. Some of these areas require further methodological work which implies that trainers be in direct contact with the field situation and that they be allowed time and resources to carry out original work. The results of this work should feedback from the sub-national level to the centre and have direct repercussions on the planning practice and procedures. They may also be expected to percolate into the curricula of universities and other higher learning institutions.

Areas for methodological work

It is realized, of course, that the approach for decentralized planning emerging from these considerations calls for training of very large and diversified audiences, for additional methodological work and for institutional and operational links between training institutions and the planning units in charge of decentralization, which are all aspects of difficult implementation and very demanding of financial and skilled human resources. **International and sub-regional training institutions as well as bilateral and multilateral lending and technical agencies** can play an important role in facilitating the above multi-faceted approach to training.

Role of international training centres

They can favour international ties, selected intercountry exchanges, the promotion of institutional and topical networks both internationally or in the form of twinning arrangements including those of institutes between developed and developing countries. They can provide financial support to allow institutions to plan and operate training in a medium-term framework and not as is often the case, in a short-term and contingent fashion. They can also give specific technical assistance and finance and promote, through specialized workshops, the training of trainers as well as consultations among professional planners and trainers on a sub-regional basis. This would facilitate the exchange and evaluation of the results achieved nationally on planning methods and procedures and on their incorporation in training programmes.

Regional and international agencies and institutions are also best suited to invest in didactic methods, materials and impact evaluation of training, areas which have not been elaborated upon in this text but which nevertheless do deserve full consideration. By their position, they can facilitate communication links at the national level that would sometimes not be possible because of institutional rilvaries; they can also help identify and support applied research and methodological work and, more generally, help the national training institutes in keeping their programmes from falling into routine, standardised teaching.

It is believed that through such concerted actions in-service training activities, as described in this text, can constitute an important contribution to the process of decentralization. The success of this depends, of course, on many other factors, human resources being however one of the most important for its driving force potential.

References

(1) FAO, Expert consultation on multi-level planning and the two-way process in agriculture and rural development planning, Bangkok, May 1983.

(2) FAO, Decentralización de la planificacion agropecuaria en America Latina (Estudio de los casos de Brasil, Colombia, Mexico, Peru y Chile), por C.A. Peixoto, 1986.

(3) FAO, Regional Decentralization for Agricultural Development Planning in the Near East and North Africa, October 1987.

(4) R. Youker "Training of African Managers: some lessons from evaluation" (paper presented at Centre for the Study of Management Learning, University of Lancaster, September 1986); and also

FAO, Methods of in-service training on agricultural project planning and analysis in Sub-Saharan Africa, March 1984.

(5) FAO, Towards Multilevel Planning for Agricultural and Rural Development in Asia and the Pacific, Paper No 52, 1985.

(6) N.D. Mutizwa-Mangiza, Local Government and Planning in Zimbabwe, TWPR, 8(2) 1986.

(7) FAO, Assessment of trained manpower needs for agricultural planning and project analysis. A case study of Gujarat, India, 1987.

FAO, "Mali: Rapport de synthèse de l'étude des besoins en formation (FORPROSA)", March, 1985.

FAO, "Burkina Faso: Identification, préparation, suivi et évaluation de projets agricoles et ruraux", April 1986.

(8) Frank Wilson. Decentralization in Kenya, the challenge to project identification and analysis methodology, p. 18, Project Planning Centre for Developing Countries, University of Bradford, U.K. (undated).

(9) Mathur, O.P. Training for Regional Development Planning: Perspectives for the Third Development Decade, UNCRD, Nagoya, 1981.

A FRAMEWORK TO ANALYSE DECENTRALIZATION

Annex 1

ITEM	NO	YES	REMARKS
1. ADMINISTRATION/ORGANISATION			
Creation of a local planning capability			
Creation of a local planning cell (pluridisciplinary)			
Decentralized planning and implementation procedures developed			
Central level high power committee to coordinate decentralization			
High calibre staff transferred to local level			
Accountability of government staff to local level authority			
Staff performance assessment made at local level			

Annex 1 Page 2.

ITEM	NO	YES	REMARKS
2. FINANCE			
Resources earmarked for decentralization			
Funds allocated to local level			
Criteria to allocate funds to local level			
Existence of un-tied funds			
Power to levy taxes given to local authority			
Incentives provided for local resources mobilization			
Financial accountability of government officers to local authority			

Annex 1 Page 3

ITEM	NO	YES	REMARKS
3. PLANNING			
Scope and content of local level planning defined			
Attempt to train and/or provide technical backstopping to local planners			
Local planning authority decides at least on location of projects			
Sectoral targets for the local level exist			
Sectoral targets prepared at local level			
Targets organized and presented in a sectoral plan (standard format)			
Existence of a data base at local level			
Existence of a local level M. & E. cell			
Possibilities of intersectoral planning			

Annex 1 Page 4.

ITEM	NO	YES	REMARKS
3. PLANNING (Cont'd.)			
Partially integrated multisectoral planning at local level			
Fully integrated multisectoral planning at local level			
Projects formulated at the local level			
Village plans formulated at village level			
Intra-district balance consideration			
Possibility to modify content and costs of projects to a certain extent during implementation without referring to the national level			

Annex 1 Page 5.

ITEM	NO	YES	REMARKS
4. PEOPLE'S PARTICIPATION			
Local planning authority includes government officials only			
Local planning authority includes a mix of government officials and local leaders			
Local planning authority includes political leaders only			
Actions to promote people's participation			
Local planning authority legally established			
Its planning prerogatives are legally established			
People participate in term of labour/finance			
People participate in management of projects			
People are consulted during the planning phase: - by survey - by participation in assemblies and forums			

TRAINING MATERIALS FOR AGRICULTURAL PLANNING

No 1 - CASE STUDY - South Nyanza Sugar Project - Kenya, 1983

No 2 - CASE STUDY - Dakawa Rice Farm Project - Tanzania, 1983

No 3 - CASE STUDY - Mkata Ranch Project - Tanzania, 1983

No 4 - Proceedings of the FAO/EADB In-Service Training Course on Project Analysis - 1983

No 5 - Note on Monitoring and Evaluation Terminology - 1983

No 6 - CASE STUDY - Ondo State Opticom Centres - Nigeria, 1983

No 7 - CASE STUDY - Waling Lift Irrigation Project - Nepal, 1983

No 8 - ETUDE DE CAS - Projet de développement de la production alimentaire en Casamance - Sénégal, 1983 (non disponible)

No 9 - CASE STUDY - Waling Lift Irrigation Project - Dasi Project Analysis - Nepal, 1983

No 10 - Schéma théorique de déroulement d'une opération de développement rural, 1983

No 11 - CASE STUDY - Credit and Marketing Project for Small-Holders in Swaziland, 1985

No 12 - Training in Policy Impact Analysis - Preliminary Plan of Action for an FAO Training Programme, 1988

No 13 - CASE STUDY - On Credit for the Wadi Arab Dam Area - Jordan, 1988

No 14 - Policy Analysis for Food and Agricultural Development : Basic Data Series and their Uses, 1988

No 15 - Structural Adjustment Programmes in Sub-Saharan Africa, 1989

No 16 - Identification and Appraisal of Small-Scale Rural Energy Projects, 1989

No 17 - Design of Monitoring and Evaluation Systems (Corum-Cankiri, Turkey), 1989

No 18 - Linkages between Policy Analysis, National Planning and Decentralized Planning for Rural Development, 1989

No 19 - Manuel de préparation des microréalisations, 1988

No 20 - Preparaçao participativa dos projectos de desenvolvimento agrícola/rural: Documento metodológico, 1988

No 21 - Rural Area Development Planning: A Review and Synthesis of Approaches, 1990 (E/F)

DOCUMENTS FOR CAPPA (Computerized system for Agricultural and Population Planning Assistance and training)

No 22	-	CAPPA Manual, 1992 (E/F/S)
No 22/1	-	The use of scenarios in agricultural sector analysis - The CAPPA system and other approaches, 1991 (E/F/S/A)
No 22/2	-	Setting targets for agricultural planning: From macroeconomic projections to commodity balances: an illustration with the CAPPA system, 1991 (E/F/S/A)
No 22/3	-	Reference international data for CAPPA applications, 1992
No 22/4	-	Projection of agricultural supply in CAPPA, 1991 (E/F/S)
No 22/5	-	A case study of the use of the CAPPA system: Cappa - Ghana, 1993
No 23/1	-	Energy for Sustainable Rural Development Projects - A Reader, 1991
23/2	-	" " " " " - Case Studies, 1991
No 24	-	Guide pour la formation de formateurs, 1991
No 25	-	Structural Adjustment and Agriculture, Report of an In-service Training Seminar for FAO Staff, 1991
No 26	-	Planification régionale du secteur agricole: Notions et techniques économiques, 1991
No 27/1	-	Rural Area Development Planning: Principles, Approaches, and Tools
27/2		of Economic Analysis. Volumes 1 and 2. 1991
No 28	-	Programmation et préparation de petites opérations de développement rural, 1992
No 29	-	Training for Decentralized Planning: Lessons from Experience, 1987 (E/F)
No 30	-	Economic Analysis of Agricultural Policies: A Basic Training Manual with Special Reference to Price Analysis, 1992 (E/F)
No 31	-	Agricultural Price Policy: Government and the Market, 1992
No 32	-	L'approche gestion des terroirs: ouvrage collectif, 1993
No 33	-	Trainer's Guide: Concepts, Principles, and Methods of Training with Special Reference to Agricultural Development, 1993
No 34	-	Guidelines on Social Analysis for Rural Area Development Planning, 1993

Copies of these materials can be requested from:

Distribution and Sales Section
FAO
Via delle Terme di Caracalla
00100 Rome, Italy

providing full details on title and number.
